30 YEARS OF MOBILE PHONES IN THE UK

Nigel Linge and Andy Sutton

To our respective wives Joanne and Steph.

First published 2015

Amberley Publishing
The Hill, Stroud
Gloucestershire, GL5 4EP

www.amberley-books.com

Copyright © Nigel Linge and Andy Sutton, 2015

The right of Nigel Linge and Andy Sutton to be identified as the Authors of this work has been asserted in accordance with the Copyrights, Designs and Patents Act 1988.

ISBN 978 1 4456 5108 8 (print)
ISBN 978 1 4456 5109 5 (ebook)

All rights reserved. No part of this book may be reprinted or reproduced or utilised in any form or by any electronic, mechanical or other means, now known or hereafter invented, including photocopying and recording, or in any information storage or retrieval system, without the permission in writing from the Publishers.

British Library Cataloguing in Publication Data.
A catalogue record for this book is available from the British Library.

Typeset in 10pt on 13pt Celeste.
Typesetting by Amberley Publishing.
Printed in the UK.

Contents

Chapter One	The First Mobiles	5
Chapter Two	Going Digital	24
Chapter Three	The Thirst for Data	42
Chapter Four	An Expensive Auction	59
Chapter Five	Smartphones	74
Chapter Six	The Broadband Multimedia Experience	86
	Acknowledgements	94

Chapter One

The First Mobiles

On New Year's Eve 1984 Michael Harrison left his family's party in Surrey and drove to Parliament Square in London so that at the stroke of midnight, as 1985 arrived, he could telephone his father, Sir Ernest Harrison. This, however, was no ordinary call, for Sir Ernest Harrison was chairman of Vodafone and when he answered, Michael said, 'Hi Dad. It's Mike. This is the first-ever call made on a UK commercial mobile network.' The UK had quietly entered the mobile era. A more public celebration followed later that same day when comedian Ernie Wise arrived at St Katherine Docks in London, in a horse-drawn Royal Mail coach, to make a mobile phone call to Vodafone's headquarters in Newbury. Seven days later on 7 January 1985, Vodafone was joined by Cellnet, who opened the UK's second mobile network.

In the thirty years that have elapsed since that momentous week in 1985, the UK has undergone a telecommunications revolution that has seen the mobile phone evolve from being a luxury item to an essential part of modern-day living and witnessed a growth in ownership that means there are now considerably more mobiles in the country than people. Today, the humble mobile phone that in 1985 was only capable of making and receiving telephone calls is now a smartphone which connects us via social media, allows us to watch television or listen to radio, provides access to the web, uses satellite navigation to guide us to where we want to be and includes a camera so that we can record and share all aspects of our daily lives.

The UK neither invented the mobile phone nor launched the world's first network, but the country has played an important role in its development. It is responsible for several firsts and is now taking a leading role in the provision of 4G networks plus helping to define the successor technology, 5G. Through this book it is our intention to reflect upon the first thirty years of the mobile within the UK by exploring how the mobile handset has changed. For many people, owning your first mobile was a liberating experience and one that is remembered with great fondness. Therefore our book will hopefully bring back many happy memories and allow the older user to wallow in nostalgia while, for the younger generation, it might prove to be a source of amazement to discover that there was a time when mobiles only made phone calls and were extremely heavy and cumbersome to carry!

The origin of the mobile phone in the UK can be traced back to the late 1940s when, in 1947, police forces began using the Pye PTC102 radio sets within their cars. Taxi firms and utility companies were similarly granted licences to operate radio systems and in the 1950s the police began using portable walkie-talkie radio sets, such as the Pye PTC122, that could be carried by officers on the beat.

A private mobile radio service employs a number of high-powered radio base stations to communicate with several mobile units within a large geographical area. The calls made on such a system are generally short in duration, with each call being allocated its own unique radio frequency. The limitations of these systems are, therefore, the number of base stations deployed, which impacts coverage, and the number of available radio frequencies – which impacts capacity. However, despite these limitations by 1971 there were 10,000 privately owned and managed radio base stations operating within the UK supporting over 100,000 mobiles. This number was growing at a rate of 17 per cent per year.

Another limitation of a private mobile radio system was the fact that they were not, and could not, be connected to the public telephone network, thereby preventing a mobile user from making or receiving conventional telephone calls. This situation changed on 28 October 1959 when the General Post Office opened the South Lancashire Radiophone Service, which not only offered integration

A Burndept BE-470 UK Police PMR transceiver (1965).

with the national telephone network but also allowed for usage by the general public. The service was extended to the London area on 5 July 1965 with the opening of the London Radiophone Service.

As the popularity of both private mobile radio and the public radiophone service grew, the ultimate limitation became the range of frequencies that were available. With each frequency supporting only one call at a time there was clearly a need for either massively increasing the number of frequencies available or developing a more efficient method for using this scarce resource.

The solution to this problem had actually been proposed back in 1947 by Bell Telephone Laboratories in the USA, who proposed a new concept of a cellular network. In contrast to a private mobile radio system, a cellular network divides the country into small geographic areas called cells that use a low-power radio base station to communicate with mobiles within that cell. Because both the radio signal range and power have been reduced, this introduces the possibility of frequency re-use. This means that so long as adjacent cells don't use the same radio frequencies, those frequencies can be re-used by other more distant cells. This immediately overcomes the limitation on the number of available frequencies, but at the expense of a more complex mobile unit that has to be able to change its operating frequencies as it moves from one cell to another.

Lynx 830FS car radiophone produced in 1975 by Dymar Electronics.

Technology in 1947 wasn't sufficiently advanced to realise this cellular concept but on 3 April 1973, Dr Martin Cooper of Motorola walked along Sixth Avenue in New York, stopped outside the Hilton hotel and telephoned Joel Engel of Bell Labs using a handheld prototype DynaTAC (Dynamic Adaptive Total Area Coverage) telephone. This event is generally recognised as being the birth of the mobile, or cell, phone. It would, however, be several years before the first trial and commercial networks opened in the USA. These delays meant that the USA fell behind other countries in developing mobile services. In Japan a cellular network opened in Tokyo in December 1979, and on 1 October 1981 the Nordic Mobile Telephone (NMT) service opened in Sweden and Norway, extending to Denmark and Finland in 1982.

In June 1982 the British government announced that two licences would be awarded to operate mobile cellular radio networks within the country, operating in the 900MHz band. These licences cost £25,000 each and required that services must commence no later than 31 March 1985 and provide coverage to 90 per cent of the UK population by 31 December 1989. One of these licences was awarded to BT in partnership with Securicor (Telecom Securicor Cellular Radio Limited), who branded their network as Cellnet, while the other was offered to open competition. In December 1982 it was confirmed that this second licence had been awarded to Racal Millicom who subsequently, on 22 March 1984, announced that their service would be known as Vodafone (voice, data, telephone). In February 1983 it was agreed that these two networks would be based on a development of the USA system and be known as Total Access Communication System (TACS).

Vodafone placed a £30 million order with Ericsson Radio Systems for the supply of the first phase of their network deployment, which was to comprise mobile exchanges in London and Birmingham together with 100 radio base stations that collectively offered capacity for 20,000 users – a figure that many thought was widely optimistic! At the time of launch on 1 January 1985, the Vodafone network comprised only ten radio base stations that provided coverage in London – this, however, was quickly expanded to cover the M1, M4 and M5 motorways and the centre of Birmingham by September 1985. Similarly, Cellnet launched their network with fifteen radio base stations serving London, but this too was soon expanded to cover Birmingham, Manchester and Liverpool.

Both the Vodafone and Cellnet networks used a variety of cell sizes ranging, typically, from a spacing of approximately 1.75 km in the centre of London up to 32 km in rural areas. This in turn had an impact on the performance of the mobile phone being used in that greater power is needed to communicate with radio base stations which are more widely spaced. In fact, there were broadly three types of mobile phone made available at this time as itemised in the following table:

Mobile phone type	Class	Power	Coverage
Car phone	Class I	10W	all areas
Car phone/Transportable	Class II	4W	most areas
Handportable	Class III	1.6W	urban areas
Handportable	Class IV	0.6W	urban areas

The car phone was intended for installation and use only within a vehicle and comprised a handset mounted on the dashboard, with all of the radio circuitry installed within the boot and an antenna on the roof. A transportable mobile was similar to the car phone in that it is intended for use within vehicles, but it was also designed with a detachable battery pack so that the mobile can be taken and used outside of the vehicle. Finally, the handportable was physically the smallest mobile device, being totally portable and intended to be carried at all times by its user. It is the handportable which has become the modern-day mobile phone, but in 1987 it represented only 15 per cent of Vodafone's subscriber base.

Vodafone initially chose two handset suppliers: Panasonic in Japan, who supplied the VM1 car phone, and Mobira (later to become Nokia) in Finland, who supplied the VT1 transportable. On the day of the launch it was the VM1 that was used by both Michael Harrison and Ernie Wise. Cellnet chose Motorola as its handset supplier. The Vodafone VM1 was a Class 1 car phone that comprised a handset and boot mounted base unit that collectively weighed 4.9 kg and retailed for £1,475.

Typical examples of the transportable class of mobile include the Motorola 4500 range which, while portable, weighed 4.2 kg and, with only 1 hour of talk time, required regular charging.

Transportables did get smaller and lighter, such as the Vodac/Panasonic EB2607, and others were commissioned for specialist usage. For example, the AA Callsafe Bag Phone was a transportable provided by the Automobile Association with connection to the Vodafone network. It was intended purely as an emergency phone to carry with you in the car and therefore provided only two dialling options to either connect you to the AA or the emergency services. A 1994 price list quotes a selling price of £149.99 with a £25 connection charge.

Left: Panasonic/Vodafone VM1.

Below: A collection of three Motorola Transportables (Partner 4800, 4500)

Vodac/Panasonic EB-2607 Class 2 Transportable.

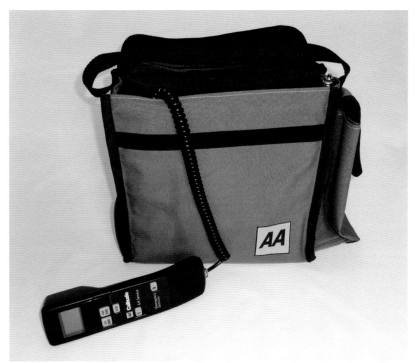

Automobile Association Callsafe Bag Transportable.

Up until the late 1980s, both Vodafone and Cellnet marketed their networks to attract the corporate or small business user who would opt for either a car phone or transportable. However, as handportables became available and prices began to fall, they began to turn their attention to the much larger consumer market. That said, prices remained high. A typical handportable could cost upwards of £3,000 to purchase, although by the late 1980s that had fallen to around £500. In addition, customers had to pay a connection charge of around £70, a monthly subscription charge of typically £30 per month and, on top of these, calls were charged at anything from 10p to 38p per minute depending on the time of day. It is perhaps not too surprising, therefore, that a common feature built into handportables was a set of timers that could monitor the duration of each call and issue audible alerts as set time intervals elapsed.

Over the thirty-year history being covered in this book, there have only been two UK companies that designed and manufactured mobile phones: Technophone and Orbitel. Technophone was set up by Nils Martensson, a Swedish engineer who had previously worked for Ericsson, in 1984. His goal was to transform the mobile phone from a large, cumbersome brick into a portable, small and usable device. Technophone sold their mobile phones through Excell Communications, branded as the Excell M1 and M2 or PC105T. The Pocketphone PC105T was released in 1986 and retailed at £1,990 but, as shown in the adverts, it did indeed fit inside a standard-sized shirt pocket! This device had a major influence on the future of mobile phone design and Technophone were soon selling 1,000 Pocketphones per month.

Perhaps the most iconic 1980s mobile phone design, one that that epitomises the decade, is Motorola's Dynamic Adaptive Total Area Coverage (DynaTAC) 8000X range. Introduced commercially in the USA in 1983, it was the culmination of fifteen years of research by Dr Martin Cooper and Rudy Krolopp and it established the mobile as a genuinely handheld and portable device. Miniaturisation had come at a price though, with talk time reduced to 30 minutes and standby time to 8 hours. The price tag of some £2,500 also placed it out of reach of the average person, so it became synonymous with the yuppie in the UK and is also affectionately known as the brick phone. A classic example of this design is the Motorola 8500X, introduced into the UK in 1987.

While Motorola dominated the early development of the mobile phone, a challenge was emerging from Finland. Nokia-Mobira Oy had been established in 1984 to produce transportables for the Nordic Mobile Telephony (NMT) network. However, the Nokia-Mobira Cityman 1320 was their first truly handheld mobile and was released in 1987 for the UK TACS network. The Motorola 888, released in 1994,

Technophone Excell M2, the mobile that truly would fit inside your pocket!

was the last and slimmest of these analogue brick phone designs to emerge from the Motorola stable.

In 1989 Motorola once again redefined the face of the mobile phone with the launch of the world's first flip phone, known as the MicroTAC. This brought major new advances in miniaturisation, style and usability, in which a moveable plate covered and revealed part of the keypad. While it may be regarded as large and heavy by today's standards, in its day it was simply amazingly light at only 290 g. Interestingly, the extending aerial was actually a piece of plastic, but was included because market research had revealed that the public expected the phone to have an external aerial. In reality the MicroTAC had an internal aerial.

Manufacturers based in the Far East continued to expand their influence. Maxon Telecom Co. Ltd was established in 1974 as Maxon Korea, with its headquarters in Seoul, and introduced the Maxon EPC590E in 1989. The Japan-based Mitsubishi group introduced the Mitsubishi MT-7 mobile phone in 1993. This particular design earned Mitsubishi three Cellnet Caesar Awards: Call Security, Features Innovation, and Clarity of User Documentation.

Left: Motorola 888, Nokia 1320 and Motorola 8500X.

Below: Motorola MicroTAC Classic.

Maxon EPC590E.

Mitsubishi MT-7.

Sony produced several novel mobile phone designs. The CM-H333 was released in 1992 and became one of the most desirable handsets of its time, retailing at £199 and being the first to come with a sliding earpiece that moved up and down to answer and end calls. It earned the nickname of the 'Mars Bar' phone because of its rather chunky design. In contrast, the Sony CM-R111 achieved further levels of miniaturisation. Released in 1993, it was distinct by virtue of having a flip-down stick microphone which Sony claimed provided a comfortable distance between ear and mouth, enabling a reduction in size and making the phone convenient to carry. However, in contrast to the vast majority of handset designs, it had no display but instead used a sequence of audible beeps and three LEDs to indicate the phone's status.

The Nippon Electric Company (NEC) released the NEC 9A in the UK in 1987 and was well received, rapidly becoming the fastest-selling mobile because it was lighter and offered more functions than its rivals. The NEC P4 was marketed as easy to hold, easy to operate and easy to carry and was introduced in 1992. It retailed for £700 and became well known for its unusual angular shape. The NEC P100 was introduced in 1993 and was a well-regarded basic phone with, for the time, a good battery life offering 100 minutes talk time and 20 hours on standby. Also known as the BT Jade on the Cellnet network, it was marketed as being light, sturdy and robust and able to fit neatly into your handbag, or jacket, so you can always have it to hand. It initially

Sony CM-H333 'Mars Bar' phone.

Sony CM-R111.

retailed for £299.99 but, for Christmas 1993, Cellnet offered a special promotion in which you could obtain it for £189.99 with 150 minutes of free calls per year included in the package.

Back in the UK, Technophone had become Europe's second-largest mobile phone manufacturer by volume behind Nokia. However, in 1991 Technophone was sold to Nokia for £34 million, which in

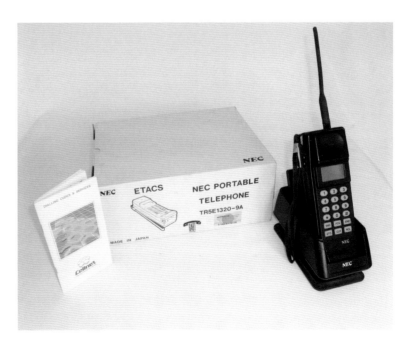

NEC 9A.

17

NEC P4.

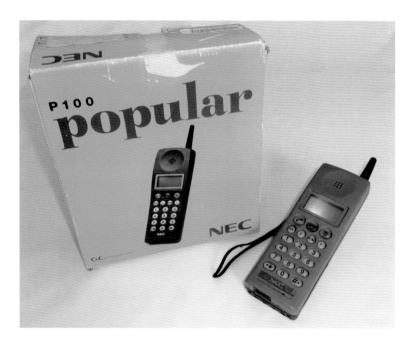

NEC P100.

turn propelled Nokia to become the world's second-largest mobile manufacturer behind Motorola. The Technophone brand did live on for a few years after the purchase; the Technophone 305 (PC215T) was released in 1992.

Technophone 305.

In 1992 Nokia released the first of their candy bar phones, the Nokia 101. The candy bar shape, or form factor, is so named because of its resemblance to a bar of chocolate and the fact that, apart from the antenna, it has no moving parts. Nokia promoted the 101 as a phone designed to slip comfortably and unobtrusively into your pocket, claiming that it was the world's most portable phone.

As the mobile phone started to become a consumer item, so the various manufacturers were faced with new challenges of marketing. In 1989, the Swedish company Ericsson rolled a bathtub onto a stage that was filled with foam bubbles and had sitting in it a man wearing a hat and holding a mobile phone. This was Harry Hotline, an adventurer who travelled around the world yet always managed to stay in touch thanks to his Ericsson Hotline mobile phone! A whole advertising campaign including television commercials, printed materials and even branded Hotline shops was devised around Harry's exploits. The Ericsson EH97 was released in 1992 and was the last to adopt the Hotline branding.

A condition of the mobile licences issued to Vodafone and Cellnet was that neither company was allowed to sell directly to customers. Instead, their services had to be offered via retailers or service providers who in turn became responsible for customer contracts, billing and setting of tariffs. This, therefore, created a whole new

Nokia 101.

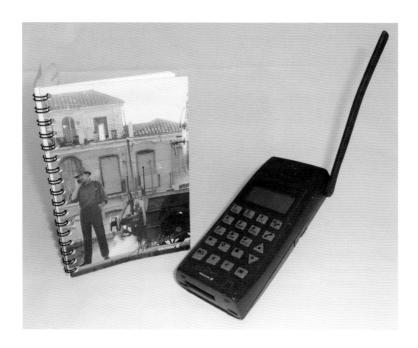

Ericsson Hotline EH-97.

commercial ecosystem with dealers and retailers appearing on the high street, each competing to offer the best deal to customers. One such company was the People's Phone, which was formed in 1988 and became the UK's largest independent mobile service provider. Their handsets, such as the PP800 released in 1996, all carried the People's Phone branding. This restriction was eventually lifted and on 19 November 1996 Vodafone bought People's Phone for £77 million.

Owning a mobile phone in the late 1980s and early 1990s was still a major event, with companies going to considerable lengths to package and present their products. By all intents and purposes the BT-branded CMH 500 of 1996, for use on Cellnet, was an unremarkable mobile. However, it came presented in its own purpose-designed plastic briefcase with mouldings to hold the phone itself, its battery and charger. Quite some distance removed from the throwaway cardboard box within which a modern-day smartphone might be packaged.

In the first ten years of the UK's mobile phone history, there was a gradual shift away from the car phone to the transportable,

People's Phone PP800.

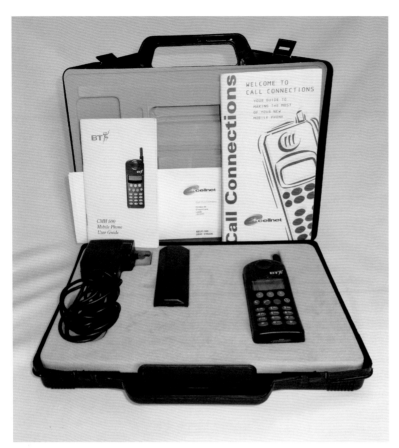

Left: BT CMH500 mobile, complete with its own plastic brief case.

Below: Motorola StarTAC 3000.

and then to the handportable. However, while the functionality of the handportable changed little, its overall appearance altered significantly, being reduced from a bulky brick phone to one that would truly sit comfortably inside a pocket or small bag. Advances in miniaturisation reached new heights when Motorola launched the StarTAC 3000 on 3 March 1996. This was not only the world's first clamshell phone but also at the time the lightest and smallest mobile ever produced. Weighing only 105 g, the StarTAC folded completely in half when not being used. Marketed as a ready-to-wear accessory but costing around £1,500 to purchase, this marvel of electronics design cost more than its equivalent weight in gold. While still only a basic featured phone, over 60 million would be sold over the product's lifetime and in 2005 *PC World* magazine rated it at number six in a list of the fifty greatest gadgets of the past fifty years.

By 1995, mobile phone ownership in the UK had reached 7 per cent of the population. The range of frequencies made available to operators had been increased to offer much greater capacity, thereby extending TACS to ETACS. However, there was one overriding limitation that was restricting the onward development of the mobile, and that was that a UK mobile phone stopped working when you reached the English Channel. The Nordic Mobile Telephone (NMT) system in Denmark, Sweden, Norway and the Netherlands, the C-Netz system in Germany, the Radiocom 2000 service in France and RTMI in Italy were all totally incompatible with each other and ETACS.

Something would have to be done about this!

Chapter Two

Going Digital

Three years before the UK launched its first analogue mobile phone network, work had already begun on it replacement. In recognition that countries were developing their own, potentially incompatible, systems and in the interests of developing a true single market, the European Commission embarked on a programme of work that would seek to develop a mobile phone standard that could harmonise mobile communication systems throughout Europe. A working group called Groupé Special Mobile (GSM) was established by the Conference of Postal and Telecommunications Administrations to realise this vision. It first met in Stockholm in December 1982 and the culmination of its work was the publication on 7 September 1987 of a Memorandum of Understanding on the Implementation of a Pan-European 900MHz Digital Cellular Mobile Telecommunications Service by 1991. Signed by telecommunication operators from thirteen countries in Copenhagen, this memorandum would later be described by Chris Gent, the then Managing Director of Vodafone, as the most important document in the history of the mobile phone. The UK played an important role in its development for it was written by Stephen Temple, a senior figure within the Department of Trade and Industry, who represented the UK's interests within the GSM work group.

While there was no preconceived intention, it was ultimately agreed that GSM would be based on digital technology using a process of both frequency and time division multiplexing. In a similar manner to the UK's ETACS network, different frequencies are used to separate one communication from another. However, transmission within a given frequency is then shared in time between a set of mobiles. All are able to transmit, but only one at a time.

The world's first GSM telephone call was made by the Finnish Prime Minister Harri Holkeri on the Radiolinja mobile network in Finland, using a Nokia 6050 car phone, on 1 July 1991. Within the UK, Vodafone was the first operator to launch a GSM service when their network went live in July 1992.

The first transportable mobile phone to be approved for use on GSM networks was the Orbitel 901, launched in May 1992. Orbitel Mobile Communications was formed in 1987 in Basingstoke and was one of only two UK-based mobile manufacturers, the other being Technophone. The Orbitel 901 certainly wasn't a small mobile and, weighing over 2 kg, had in many ways regressed to the early days of the analogue brick. Of course there was a good reason for this in that

GSM was completely different to first-generation analogue networks, thereby necessitating the development of totally new electronics.

However, the Orbitel 901 has another claim to fame – it was the mobile on which the world's first text message was received. The Short Message Service (SMS) was developed as a secondary service for mobile operators to be able to send network update messages to its customers. Being limited to 160 characters per message, it was never conceived that it might one day become a primary form of communication for millions and stimulate the texting revolution. So it was on 3 December 1992 that Neil Papworth used his computer to compose and send a SMS text message to his Vodafone colleague, Richard Jarvis, who received it on his Orbitel 901. Given the intended purpose of SMS, the first-generation GSM mobiles were designed only to receive such messages and this one simply read 'Merry Christmas'.

Motorola launched the world's first GSM handportable mobile, the Motorola International 3200, on 3 September 1991. Resembling one of the earlier analogue phones from Motorola's 8000 range, the 3200 actually became the UK's most popular GSM mobile, retailing for

Orbitel 901, the world's first GSM mobile.

Motorola 3200 – the world's first GSM handportable.

£762 at Carphone Warehouse in August 1993 but dropping to £199 by the end of that year.

Nokia's first GSM handportable was the Nokia 1011, so called because it was launched on 10 November 1992. Also known as the Nokia Mobira Cityman 2000, the launch of the Nokia 1011 helped propel Nokia on its way to becoming the world's largest mobile manufacturer. Having decided to back GSM as the future, Nokia was able to design and produce mobiles that could immediately be sold and used throughout Europe. Motorola, on the other hand, continued to invest and develop their analogue ranges. While at the time the Motorola mobiles may well have been the better quality phones, each analogue market was different, necessitating a slightly different design of handset. This would ultimately hold Motorola back and, in 1998, Nokia was able to overtake them as the world's number one, a position Motorola would never reclaim. Indeed, today the pioneer of the mobile phone seems to have all but disappeared.

The UK government was keen to encourage further competition within the mobile market and to expand capacity. It therefore commissioned the Department of Trade and Industry to produce a consultative document called 'Phones on the Move'. Published in January 1989, this document set forward plans to licence new operators who could provide GSM networks operating at 1800MHz,

Nokia 1011, the mobile that would help propel Nokia to become the world's largest mobile manufacturer.

which would be known as Personal Communication Networks. Lord Young, Secretary of State for Trade and Industry, announced that one licence would be offered to Cable & Wireless, who operated Mercury Communications, while a second and third would be opened up to bidders. This competition was subsequently won by Unitel, which was owned by a consortium including US West, and Microtel, which was owned by a consortium that included British Aerospace. In 1992, however, Cable & Wireless and US West merged to form Mercury Personal Communications, through which they launched Mercury one2one. British Aerospace sold Microtel to Hong Kong-based Hutchison Telecom. As a consequence, the original three PCN licences became two – meaning that the total number of UK mobile operators had grown to four.

The Mercury one2one network became the world's first 1800MHz GSM network when it was launched on 7 September 1993. Initially the network was restricted to the London area, but offered a unique proposition to customers: free calls during the weekday evenings (7 p.m. – 7 a.m.) and throughout the weekends. The first two mobiles made available to customers were the entry level Siemens m200, retailing at £249.99, and the higher end Motorola m300, which cost £299.99. By the end of 1994 Mercury one2one had over

Siemens m200, the world's first 1800MHz GSM mobile.

Motorola m300, Mercury one2one 1800MHz GSM launch mobile.

200,000 connections and had extended its reach out of London into Birmingham and the West Midlands.

The UK's third GSM network to launch was that from Cellnet, which opened in December 1993, and this was followed by the fourth, launched by Hutchison Telecom on 28 April 1994; the second of the two new entrants at 1800MHz. Hutchison's approach was to focus on providing national coverage with products targeted at the general consumer. They named their network Orange and launched it with the catchy slogan 'The future's bright, the future's Orange'. Uniquely at that time, per-second billing, caller-ID and talk plans with inclusive minutes were offered as standard. Two mobiles were made available at launch and these were the Nokia 2140, which was the smallest GSM1800 mobile of its time, and the Motorola mr1. By the end of 1994 Orange had attracted 379,000 customers, a figure which more than doubled the following year.

In developing the GSM standard, the way in which a mobile phone is identified on the network also changed. Whereas with first-generation analogue mobiles the telephone number was hard-wired into the phone itself, with GSM this information was contained in a removable

Nokia 2140 Orange launch mobile.

Motorola mr1 Orange launch mobile.

Subscriber Identity Module, or SIM, card. This allowed manufacturers to mass produce phones which could then be configured for a particular operator in a given country by inserting the correct SIM. Initially these were the size of a credit card, but as the GSM mobile was developed and made smaller, a miniature version became standard.

The transition to GSM created a mass market, which in turn helped to drive handset prices down and move the mobile from being the preserve of the business user to become a must-have device for the general consumer. Mobile phones would soon be seen as an essential part of modern living and 1999 was a tipping point for the UK – a new mobile phone was being sold at the rate of one every 4 seconds and ownership doubled to 46 per cent of the population, and then reached 73 per cent the year after. This massive expansion of the market brought new challenges for the mobile manufacturers, who had to find ways of encouraging people not only to buy their first device but their second, third and fourth. Companies therefore started to develop and market the mobile phone as a device that combined the convenience of being able to keep in touch on the move with a fashion item and entertainment centre. Among these, Nokia was one that truly embraced and exploited the mobile as a device that could be personalised and, most importantly, be specifically targeted at the younger customer.

Examples of large and small SIM cards.

Released in 1994, the Nokia 2110 became the first to use the now famous, indeed iconic, Nokia ringtone based on a guitar work named 'Gran Vals' by Spanish musician Francisco Tárrega, written in 1902. Offering customers a range of ringtones became common place, including the ability to compose your own using tones selected from the keypad. This in turn spawned a whole new industry that made its money from selling ringtones that could be downloaded. In the history of the ringtone there is one that most definitely stands out from the crowd by some considerable margin. It began life in 1997 when a Swede called Daniel Malmedahl decided to record himself imitating the sound of a motorbike engine. Six years later, another Swede, Erik Wernquist, took Daniel's soundtrack and used it as the basis of a 3D animated character which he created and called 'The Annoying Thing'. The ringtone company Jamster bought the rights to use it as a ringtone in 2004 and launched it onto the world as the 'Crazy Frog', which went on to become the bestselling ringtone of all time, generating almost half a billion dollars in revenue over its lifetime.

Attention was also paid to the mobile's design and ease of use. In 1996 Nokia launched their first slide mobile, the Nokia 8110, which had a sliding cover protecting the keypad. This could be moved downwards to give access to the keys and to bring the microphone closer to your mouth. Nokia claimed that this was the first mobile to be uniquely designed to fit the contour of the human face. It was,

Nokia 2110i, the first with the Nokia tune ringtone.

Examples of the Nokia 8110.

however, a modified version of this model that found fame when it was used by Neo in the 1999 feature film *The Matrix*.

From a user's point of view, as mobiles acquired more functions so being able to select options from menus became more cumbersome. Nokia started to address this issue with the Nokia 3110, which was announced in 1997. This was the first model to feature their Navi-Key menu navigation system, which Nokia claimed offered the ultimate ease of use with a unique one-key access to functions.

In 1997 Nokia decided that mobiles should be fun and introduced their iconic game of *Snake* into the Nokia 6110. This obsessive game required you to control a pixelated snake as it moved around the screen, feeding it to make it grow bigger but all the time ensuring that it never caught its own tail! This game proved incredibly popular, being loaded onto millions of mobiles, and was so successful that it kick-started the whole mobile gaming business which continues to thrive to this day.

Personalisation was extended beyond your choice of ringtone in 1998 when Nokia launched the 5110, which had interchangeable Xpress-on covers. The Nokia marketing campaign stated that now you could now match your Nokia mobile phone with almost any clothing, style and occasion. Initially a simple colour range was available, but

Nokia 3110, the first to use the Navi-Key menu navigation system.

Nokia 6110, the first to feature the iconic game of *Snake*.

the concept soon grew to become a standard feature of mobiles with a major business emerging to support the production of an enormous range of general, special and limited edition covers.

With the younger customer firmly in their sights, Nokia launched the 3210 in 1999, which became one of the most popular and successful mobiles ever produced with over 160 million being sold. It was replaced in 2000 with the equally iconic Nokia 3310, which sold 126 million, and then in 2002 with the Nokia 3410, which was Nokia's first Java-enabled mobile.

Finally, the Nokia 5510 launched in 2001 was most definitely first and foremost an entertainment device and a phone second. Advertised as looking weird but sounding great, the 5510 could store up to 2 hours of music and included an FM radio.

Originally conceived as a means of unifying mobile services within Europe, GSM has gone on to become a world-beating technology. However, while in the UK GSM services are provided on both 900MHz and 1800MHz, the USA offers services at 1900MHz. Bridging this gap between Europe and the USA was therefore hugely important

Nokia 5110 with Xpress-on interchangeable covers.

Nokia 3210, 3310 and 3410.

Nokia 5510.

for establishing GSM as a universal standard. The Motorola mr601, launched in 1997, offered dual-band coverage at 900MHz and 1800MHz, thereby enabling it to be used on any of the UK's four networks. However, the Bosch World 718 was the first mobile to offer global coverage when it was launched in 1998 supporting both 900MHz and 1900MHz. This meant that either a Vodafone or Cellnet customer could use the same device in the UK and USA. However, it was the Motorola Timeport L7089 released in 2000 that provided tri-band coverage on all three frequencies and hence on all UK and USA GSM networks.

With the mobile becoming a mass-market consumer item and ownership exceeding half of the population, it was now that text messaging truly took off. In 2000, 6.2 billion text messages were sent over the UK GSM networks. Large though this figure was, text message usage would continue to grow rapidly until it reached a staggering 155 billion per year before beginning to fall as a result of users switching to social media applications. However, in the process, text messaging redefined our language and introduced abbreviations such LOL and OMG into common usage. This quite unexpected growth caught the mobile industry out. The first generation of GSM mobiles could only

Motorola mr601, dual-band (900MHz and 1800MHz).

Bosch World 718, dual band (900MHz and 1900MHz).

Motorola Timeport L7089, tri-band (900MHz, 1800MHz, 1900MHz).

receive SMS and the design of a standard keypad was not well suited to someone wishing to type a message. A solution to this came through the development of predictive text, where letters of the alphabet could be typed by pressing the numeric keys several times with software then trying to automatically complete the word you were typing. Text on 9 keys, or T9 predictive text, was developed by Tegic Communications and first used by Nokia on their 3210 mobile in 1999. Motorola developed an alternative called iTAP, which was also first used in 1999 on their L7089, and Blackberry introduced SureType in 2004 onto their 7100 range of mobiles which, unusually for Blackberry, did not have a traditional QWERTY keypad. Companies such as Motorola even produced devices like the v100 that were designed specifically for texting, so much so that it didn't have an in-built speaker, requiring you to use a headset when making phone calls.

Motorola v100 Personal Communicator.

The increased demand and competition that existed within the UK market also generated new approaches to charging for services. In August 1997 One2One (rebranded from Mercury one2one) became the first UK mobile phone operator to offer a pay-as-you-go service. Needless to say, all of the other mobile phone operators followed suit with their equivalent service: for example, Vodafone's Pay As You Talk. Some companies were even prepared to give the mobile phone away for free as an incentive to use their service which, given the cost of the first GSM mobiles, would have been unimaginable a few years earlier.

On 1 October 2000 BT Cellnet (rebranded from Cellnet in 1999) closed down their analogue ETACS network and Vodafone closed theirs on 31 May 2001, at which point the UK became a fully digital mobile country and the sixteen-year era of its analogue mobiles ended. Moving to GSM had, however, dramatically redefined the marketplace, with the mobile becoming a common everyday object, affordable to millions and beginning to be differentiated by its non-voice features. Size was also important with mobiles shrinking to some of the smallest form factors ever seen – the Innox i-800, for example, claimed to be the world's smallest mobile measuring only 70 x 41 x 22 mm.

A Vodafone MN-1 available on their Pay As You Talk service.

A Philips TCD308 Diga GSM mobile being given away for free as a marketing incentive.

Innox i-800.

Nokia 1100, the world's top-selling mobile phone.

Manufacturers were able to exploit the huge market which GSM created and push their sales to new heights. The world's top-selling mobile of all time is the Nokia 1100, which was launched on 27 August 2003 and achieved sales of 250 million worldwide.

With the dawning of the new millennium, however, the onward development of the mobile phone would be firmly driven by a desire to increase and expand its non-voice capability and, specifically, to fully integrated it with the Internet.

Chapter Three
The Thirst for Data

The first step towards mass-market mobile data communications was taken with the introduction of the Wireless Application Protocol (WAP). WAP services could operate over traditional GSM and SMS, and this was how early services were launched. The growing adoption of home PCs and broadband generated great demand for Internet access on the move and in particular on the mobile phone itself, rather than via a laptop with a data card. WAP was developed through the collaborative efforts of many within the mobile ecosystem, the WAP Forum launched in 1997 with key inputs from major players such as Ericsson, Motorola and Nokia. This resulted in a series of technical specifications which defined file format (Wireless Mark-up Language or WML) and architecture, including WAP gateways. This created a technology which supported a wide range of applications from web browsing to messaging, while the wide industry support ensured compliance to the specifications and therefore the all-important interoperability among vendors. The basic WAP Forum specification was released in April 1998 with the final version of WAP, known as WAP2.0, being released in July 2001. As mobile phones evolved, particularly in terms of processing power and full-colour screens, developers started to produce standard web content which was supported in parallel with WAP and ultimately replaced it; one of the last WAP sites to be switched of was that of the BBC in January 2011.

The Nokia 7110 was announced on 23 February 1999 at the GSM World Congress in Cannes, France. In a press release issued by Nokia, they claimed that the 7110 was the world's first media phone based on WAP. It followed on from the Nokia 8110, and at the top, on the rear of the phone, is a small silver button which when pressed releases the slider, thereby revealing the keypad and also answering a call. Another innovative design feature was the NaviRoller, which is a thumbwheel that can be seen just below the screen in the middle of the phone. This provided easy access to menus, allowing the user to scroll through items by rotating the roller and then selecting the required item by pressing the roller.

Unfortunately WAP was somewhat over-sold in those early days, slogans such as 'Surf the BT Cellnet' suggested the Internet experience would be comparable with that experienced by many on home PCs with dial-up or early broadband ADSL access. This was certainly not the case.

The growing desire for mobile Internet drove the development of the General Packet Radio Service (GPRS), which was added to the GSM

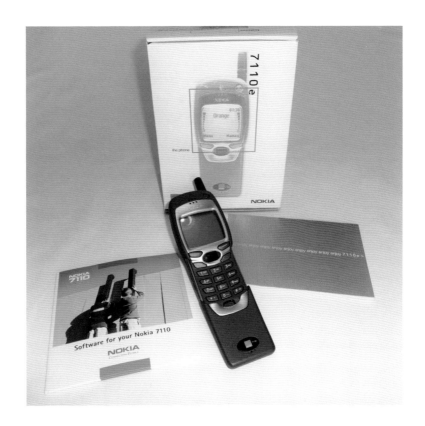

Nokia 7110 WAP phone.

Motorola Timeport with WAP capability.

specifications in 1997 to provide enhanced data communications. From the perspective of mobile network operators, GPRS required a significant investment in new network infrastructure and a need for new skills which were realised through retraining existing staff and recruiting new staff from a data-networking background who would work alongside the cellular engineers. The foundation laid in those early GPRS standards is still with us today in the latest 4G mobile networks; the data rates, latency and other key parameters have improved considerably of course, but many of the protocols are similar to or evolutions of those original GPRS specifications.

GPRS fundamentally changed GSM. While the radio interface was effectively the same, the way it operated was quite different. GPRS introduced four data transmission schemes to the radio interface which had to be supported across the transmission network that connected the base station to the new Packet Control Unit (PCU), as shown in the network architecture diagram. The introduction of these four schemes was the start of a phenomenon that's still with us today; as you move away from the cell site, the user data rate reduces because the radio interface needs to maintain a robust error-free connection. GPRS could support data transmission rates of some tens of kilobits per second.

GPRS opened up the possibility of mass-market mobile Internet access, however in reality it would be some years before what we know as the rich multimedia Internet of today was available on a mobile phone. Initial consumer offerings were quite limited in functionality, often text-based information services such as news and sports results. GPRS was also exploited by corporate organisations to mobilise their workforce by allowing access to email, corporate information and databases on the move, thereby laying the foundations of today's mobile working environment.

The original GSM network was effectively a traditional telephony network with mobile radio access that supported basic data services, much the same as the home telephone line supported dial-up connection to the Internet prior to broadband. The introduction of GPRS changed the radio access network, however it wasn't too obvious as most of this was implemented via software updates. The most significant change for network operators occurred within the core network with a completely new parallel network designed to support packet data. This new core had its own interface from the access network and connected directly to external data networks such as the public Internet and private corporate networks. The traditional telephony network, also known as the circuit-switched network, was connected to the new packet-switched network to enable subscriber authentication and location information to be shared.

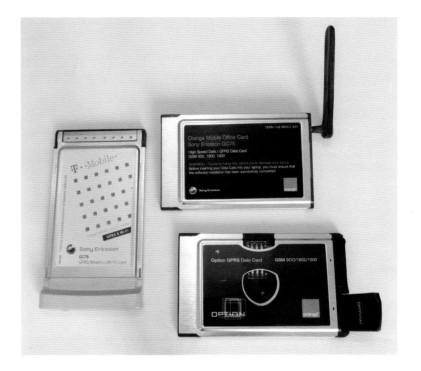

GPRS data cards.

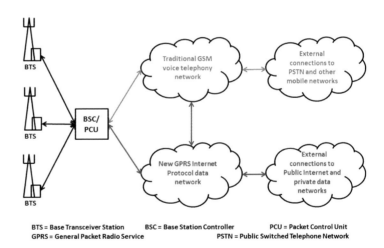

BTS = Base Transceiver Station   BSC = Base Station Controller   PCU = Packet Control Unit
GPRS = General Packet Radio Service   PSTN = Public Switched Telephone Network

GSM/GPRS network architecture.

BT Cellnet launched the UK's, and world's, first GPRS service in June 2000. In the UK this was followed by Vodafone in April 2001, Orange in December 2001 and T-Mobile (rebranded from One2One by this time) in June 2002. The provision of this always-on best effort data service (commonly known as 2.5G) continued to fuel the growing consumption of data services. Once a network operator had deployed

GPRS, they had built the underlying infrastructure to support future mobile data services, such as the Internet, and remote access to corporate data became an increasingly important service. When GPRS was launched approximately 25 per cent of the UK population had Internet access, by 2015 this had grown to approximately 90 per cent. Nowadays, many people access the Internet daily via mobile devices and, for some, mobile is their only means of Internet access.

Early GPRS handsets included: Ericsson R520m, Motorola P7389i and T-260, Nokia 8310, Samsung SGH-Q100 and Siemens S45i. BT Cellnet launched their service with the Motorola T-260, which cost £199 at launch. A subscription to the GPRS service was an additional £7.99 per month on top of the chosen voice calls tariff and included 1MB of data, with consumption above the 1MB limit attracting an additional charge of £3.99 per MB. Alternatively, users could opt for a data package costing £3.99 a month plus two pence for each kilobyte of data consumed.

Mobile phone manufacturers started to market devices at particular segments. The business users required high-end devices and in 2002 Nokia launched a winner; the Nokia 6310 became an incredibly popular phone with business users. The device continued to sell well until it was withdrawn in late 2005.

The development of GPRS was eagerly embraced by business users in the first instance and the early email-optimised Blackberry devices became the mobile business person's constant companion. Blackberry

Ericsson R520m.

Motorola Timeport T260.

Nokia 8310.

Siemens S45i.

Nokia 6310i.

is owned by Canadian company Research In Motion (RIM), which was formed by two engineering students, Mike Lazaridis and Douglas Fregin, in 1984. The company originally focused on packet-switched data communications for the emergency services and security markets, however much of their research was to come together to create a mobile phone ecosystem like nothing else before it.

Announced in 2001, the Blackberry 5820 was the first cellular Blackberry device to be launched in the UK. The device operated in the 900 and 1800MHz GSM/GPRS frequency bands, however it required an external headset to make and receive phone calls as it didn't have a built-in microphone or speaker. Later models would be built to operate in the more conventional mobile phone mode. This device was marketed as a mobile phone with email functionality, whereas previous Blackberry devices had been known as pagers (albeit two-way devices). BT Cellnet was the first UK mobile network operator to support the Blackberry service, but others quickly followed. During the summer of 2003 Blackberry launched the 7230 device, which was their first handset with a colour screen. Blackberry started to dominate the business market, but the traditional mobile phone manufacturers still dominated the consumer market. It would be a few years before Blackberry attempted to enter this market however, when they did, the take up was phenomenal. Spurred on by free Blackberry Messaging, or BBM as it was commonly known, a huge new market grew up very quickly among the tech-savvy youth to whom instant communications was becoming the norm.

Another benefit of developing the data capability of the mobile was the ability to start exploiting multimedia communications. None more

Blackberry 5820.

Blackberry 7230.

so than the development of the camera phone. The world's first true camera phone was the Sharp J-SH04, this contained a camera which was fully integrated with the phone and enabled the user to send photographs directly from the handset. This device was released in Japan during 2001.

The UK, like the rest of Europe, was somewhat slower in adopting the camera phone. Initial services were based on MMS (Multi-Media Messaging), or 'picture messaging' as it became known. T-Mobile launched the UK's first picture-messaging service in June 2002, however the only phone available to support this new service was the Sony Ericsson T68i, which one could argue was not a true camera phone given the comparative definition of the Sharp J-SH04. The Sony Ericsson T68i came with a camera attachment which connected to the base of the phone to enable operation, and the proposition was quite expensive too. Once the handset and camera attachment had been purchased there was an additional £20 per month charge on top of the subscriber's standard calls and texts plan. Orange launched MMS later in 2002 and adopted a very different go-to-market strategy – rather than charging an additional monthly fee they charged on a per message basis. Although not the first mobile phone with a colour screen, it's fair to say that the Sony Ericsson T68i was the UK's first mass-market mobile phone with a colour screen.

Sony Ericsson T68i with camera attachment.

Sony Ericsson T68i with camera attachment fitted.

The Sony Ericsson T68i has another claim to fame: it was the first device to be branded 'Sony Ericsson' after the creation of the joint venture between Ericsson and Sony in 2001. The Sony Ericsson joint venture lasted until 2012 when Sony bought Ericsson's share of the business. During the lifespan of Sony Ericsson some iconic

Sony Ericsson W810i.

Sony Ericsson W580i 'Walkman' phone.

brands appeared on mobile phones. The Sony Ericsson W800 was the first phone with the 'Walkman' branding as music became truly integrated with the phone in 2005, others followed including the W810i and W580i. The Sony 'Cybershot' branding (still used today on

Motorola C550 camera phone.

Sony digital cameras) highlighted the arrival of high-quality camera phones. Early Motorola camera phones included the E365 and C550.

By late 2002 all four UK operators offered a commercial MMS service with Nokia's latest device, the Nokia 7650, which had a fully integrated camera. In addition to the built-in camera the 7650 came pre-loaded with the necessary software to make taking, managing and sending photos easy. This software was based on Nokia's Symbian OS. The Nokia 7650 quickly established itself as the bestselling camera phone of its time and with the growing MMS market came ever lower picture-messaging prices and therefore adoption of this technology became the norm. The Nokia 7650 had a 0.3 megapixel rear-facing camera. Nokia's first 1 megapixel camera was the 7610, launched in 2004. Sharp produced a number of early camera phones. Their first for the European market was the GX10, which had 110,000 pixels along with a digital zoom capability. Due to its innovative features the GX10 was voted 'Best Handset' at the 3GSM World Congress in 2003.

Motorola pioneered the clamshell mobile phone design when they launched the analogue StarTAC in 1996; at the time it was the world's smallest and lightest mobile. This was followed by a GSM version in 1997 and, in 2004, they surpassed all previous designs with the launch of the V3 Razr. This was an ultra-slim clamshell mobile measuring only 13.9 mm thick thanks to its revolutionary precision-cut illuminated keyboard in which the numbers and letters were chemically-etched.

Nokia 7650.

Sharp GX10.

As demand for faster data rates increased and ever higher volumes of data were transmitted over mobile network, it became obvious that GSM with GPRS was not going to scale sufficiently to meet future demand. This future demand had been predicated by the European Telecommunications Standards Institute (ETSI) who, in

A closed Sharp GX10, showing the digital camera.

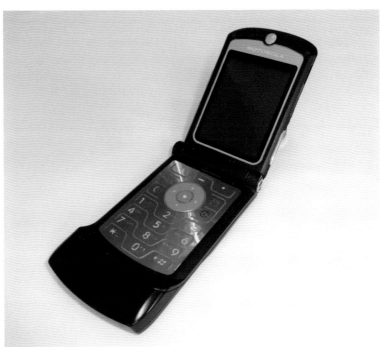

Motorola V3 RAZR.

1991, established a group known as SMG5 (Special Mobile Group 5) to develop a new standard that would become known as UMTS. 3G, as it is commonly known, was however somewhat delayed due to its cost and complexity, so in parallel with this an evolution of the GPRS radio interface was undertaken. The evolved radio interface was known as EDGE and increased the data rates and therefore system capacity by a factor of three compared with basic GPRS. The original meaning of EDGE was 'Enhanced Data rates for GSM Evolution', however given the increasing popularity of GSM outside of Europe a second interpretation is found: Enhanced Data rates for Global Evolution. The increased radio interface capability was achieved through the implementation of a new modulation scheme and range of transmission schemes on the radio interface which offered the potential for connections exceeding 100kbps, quite remarkable at the time! Orange was first to launch EDGE capability in the UK during February 2006, which was some time after the standardisation of the technology and EDGE support was already available in many 2G handsets. Orange launch devices typically supported EDGE and 3G with examples including the Nokia 6280 and N70. Additionally, Orange launched the first Blackberry EDGE device, the Blackberry 8700, and added EGDE capability to its SPV range of devices.

GPRS and EDGE capability was to be truly tested as a result of Steve Jobs unveiling the Apple iPhone on 9 January 2007 at the Macworld 2007 convention in San Francisco. This first-generation iPhone was officially launched on 29 June 2007 when many network operators had deployed their 3G mobile networks. Despite this, a decision was taken to build the first device as a 2G-only smartphone. It operated

Nokia 6230 2G phone with EDGE.

EDGE-capable Blackberry 8700.

Orange SPV E610 with EDGE.

on four GSM bands (850MHz, 900MHz, 1800MHz and 1900MHz) which allowed Apple to address a significant global market with data services able to operate on GPRS and, if available, EDGE technology. As we'll discuss in the next chapter, 3G technology was slow out of the starting blocks for several reasons, and due to the frequency band in use would lag 2G coverage for many years. Therefore, launching a 2G device was a strategic decision. The iPhone first appeared in the UK during November 2007 on the O2 (rebranding of BT Cellnet) network.

The ownership of an iPhone drove consumers to increase their consumption of data services, which caused significant challenges for network operators. This lead to the deployment of new 'in-fill' cellular capacity base stations and/or implementation of EDGE capability to manage subscribers' rising expectations of mobile data performance. The go-to market strategy adopted by Apple was very different to anything seen before, certainly in the UK, with an exclusive agreement with a single network operator; therefore if you wanted an Apple iPhone, you went to O2. After two years of the exclusivity deal it was announced that Orange would start to sell the iPhone 3G and 3GS during late 2009. Vodafone completed a deal announced two days after the Orange press release, which allowed them to start selling the iPhone in January 2010. Deals with others operators followed and nowadays all UK operators support Apple iPhone and other Apple products on their networks. To manage this massive explosion in data demand, the mobile industry had to respond with new network capability – this came in the form of 3G.

Chapter Four

An Expensive Auction

The 3G story, like 2G before it, started prior to the commercial launch of its predecessor. The European Telecommunications Standards Institute (ETSI) established a Special Mobile Group, known as SMG5, in 1991 to explore the standardisation of a third-generation mobile communications system to be known as the Universal Mobile Telecommunications System (UMTS). As part of an agreed global radio frequency assignment for UMTS, the 2100MHz band was identified for 3G use in the UK. The initial release of the 3G standards, known as Release 99, was published in 1999, however it was not until the year 2000 that the UK held an auction for 3G spectrum in the 2100MHz band. The spectrum auction was managed by the Radio Communications Agency (now an integral part of Ofcom) and commenced on 6 March 2000. The use of an auction as a technique to allocate radio spectrum was new to the UK. 2G licences had been awarded based on an assessment of the quality of proposals received from interested parties, as had been the case with the original licences awarded to Cellnet and Vodafone to enable the operation of the analogue TACS networks. The Radio Communications Agency described the reason for this new approach as follows: 'Auctions are a fast, transparent, fair and economically efficient way of allocating the scarce resource of radio spectrum. Government should not be trying to judge who will be innovative and successful.'

To ensure the UK mobile communications' ecosystem continued to evolve in a competitive manner it was decided that a new entrant, a fifth network operator, would be enabled as a result of the auction process and therefore an amount of spectrum was reserved for this new entrant. A total of thirteen companies applied to participate in the auction and these included the four existing mobile network operators along with nine potential new entrants. The spectrum to be auctioned was split into five lots referred to as Licence A through to Licence E, Licence A being reserved for the new entrant. The amount of spectrum allocated to each license would not be the same, therefore operators had to make strategic decisions as to which licence they wanted to bid for. The largest spectrum allocation was reserved for the new entrant to compensate for their lack of 2G spectrum.

The auction website provided daily updates on the current bids for particular licences; people watched in amazement as the values increased day on day, all forecasts and budgets disappeared as bidders competed to be a part of the 3G future. After 150 rounds of bidding the final five participants were left, these were the licence winners. The original estimate within government circles was that the auction would raise £5 billion, a huge sum of money considering the value of previous spectrum awards. The reality was that the final amount bid by the five winners was a staggering £22,477,400,000 distributed across the five licences, as shown in the following table.

Licence A: TIW at a cost of £4,384,700,000
Licence B: Vodafone at a cost of £5,964,000,000
Licence C: BT at a cost of £4,030,100,000
Licence D: One2One at a cost of £4,003,600,000
Licence E: Orange at a cost of £4,095,000,000

The new entrant was Telesystem International Wireless (TIW) UMTS (UK) Limited, a subsidiary of TIW, the North American telecoms company, with backing from Hong Kong-based Hutchison Whampoa. TIW had formed a joint venture with Hutchison Whampoa to build out and deploy the 3G network, however it wasn't long before Hutchison bought out TIW completely and therefore owned 100 per cent of what was to become known as H3G UK Limited and marketed as '3' (Three).

By the time the 3G spectrum was awarded, the four established mobile network operators had attained a good level of UK population and geographical coverage with their GSM networks; competing with this would clearly be a barrier to market entry for the new entrant, therefore it was decided that an existing operator must provide access for national roaming until such a time as the new entrant was happy with the level of coverage they had on their own network. The initial national roaming contract was signed with BT Cellnet prior to Three's commercial launch. In 2006 the national roaming contract switched to Orange after H3G had launched a commercial tender for the renewal.

The first network to launch a 3G service on the UK mainland was Hutchison with their new Three brand. Commercial service was launched on 3 March 2003 (3/3/3), a once-in-a-lifetime date and therefore an opportunity Three had to be ready to take! How much network was actually available on this date is an academic point as such – 3G had arrived and the other operators had to step up to compete with the new entrant.

The first public 3G call was made by Trade and Industry Secretary Patricia Hewitt, who called Stephen Timms, Minister for e-Commerce. As it turned out, the supply of handsets was limited anyway and therefore the consumer offer available on launch day was to pre-order handsets for delivery later in March. Three launched with three different handsets, however the devices were rather bulky and suffered from poor battery life, especially when compared with the state-of-the-art 2G phones of the day, which now supported colour screens, cameras and GPRS – possibly EDGE too. The Three launch phones were Motorola A830 at £125, NEC e606 at £220 and NEC e808 at £225.

Each of these phones was subsidised by Three to enable them to compete with the established operators. Monthly subscriptions were expensive and although the other operators launched during 2004 and 2005, it's fair to say that early 3G didn't capture the public's imagination initially. As devices improved and the differential from 2G started to become obvious, volumes started to increase. By August 2004, Three had connected a million customers. The LG U8110 is an example of the smaller and more feature-packed handsets that Three brought to market and launched during 2005. The Nokia N70 hit the market in 2005 too and became a 3G bestseller.

Vodafone was the first of the four established mobile operators to launch their 3G service in April 2004, followed by Orange in July 2004, O2 in February 2005 and finally T-Mobile in October 2005.

Three launch phone Motorola A830.

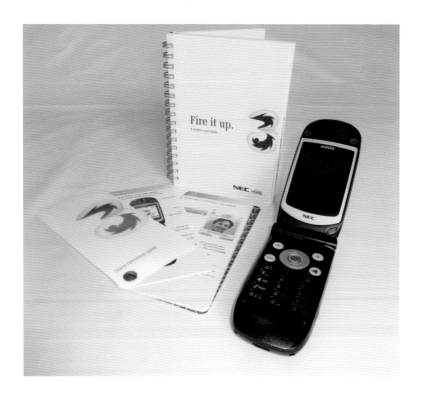

Three launch phone NEC e606.

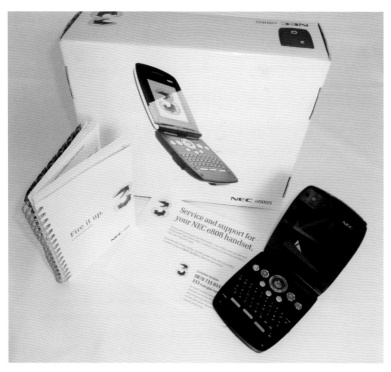

Three launch phone NEC e808.

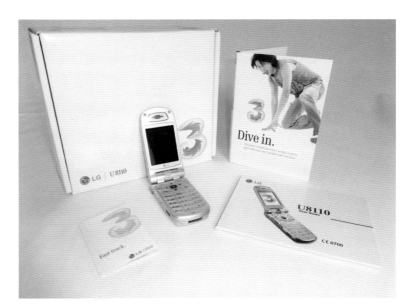

LG U8110.

Nokia N70.

While 3G handsets took a while to gain traction and public acceptance, the same cannot be said for 3G data cards. The 3G networks launch capability was 384kbps on the downlink (network to device) and 64kbps on the uplink (device to network). Given the asymmetrical nature of web traffic and the fact that both connections were significantly faster than any of the 2G technologies achieved in reality, business users started to adopt 3G data and data cards shipped in volume from late 2004.

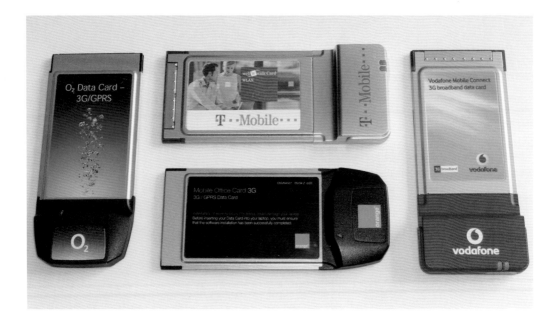

3G data cards.

The higher data rates of 3G are enabled by the use of a radio technology known as Wideband Code Division Multiple Access (WCDMA) along with higher speed connections and broadband network components within the packet data core networks. Much of the early publicity around 3G focused on data rates of 2Mbps. However, in reality, this was the total capacity of a single WCDMA carrier at launch and was therefore never going to be available for a single user, however 384kbps was mighty impressive at the time!

The Nokia 6650 was the first phone to support WCDMA in the 2100MHz band and announced by Nokia late in 2002 and commercially launched in 2003. This device was used extensively by UK mobile network operators to test their 3G networks but was never launched as a consumer product in the UK.

The Nokia 7600 adopted a completely new form factor, one which wouldn't catch on as such. Nonetheless it illustrated the ambition and vision of 3G for evolution beyond the simple handheld mobile telephone. Three launched the Nokia 7600 in February 2004 but, due to its lack of a forward-facing camera, it didn't support video calling. This was the first Three device not to offer this service. Mobile video calling didn't take off in the way it had been anticipated and the killer app for 3G would become mobile data and Internet access.

Vodafone launched 3G in the UK with six handset variants including the Sharp 902, which was claimed to be Europe's first mobile phone with a 2 megapixel camera phone.

Nokia 6650.

Nokia 7600.

Nokia 7600 – making a fashion statement.

Orange 3G handsets were available from £199.99 with contracts available from £30 per month. Orange claimed 70 per cent 3G population coverage at launch. As an aside, Orange Wednesdays was launched in the UK during 2004, giving Orange customers two-for-one cinema tickets for any film (on a Wednesday, of course) and this promotion ran until early 2015.

In 2005 Orange launched its Signature devices for business, including the SPV M5000, the first 3G device designed specifically for the business market with a new form factor that offered a real alternative to a laptop computer.

3G handsets had, to date, been quite large when compared with the smaller 2G-only phones, however smaller 3G phones soon became available. Sony Ericsson announced the K610i in February 2006 and claimed it was the smallest and lightest 3G mobile phone on the market. Evolution of camera phones occurred at a remarkable pace in the 3G era – the Sony Ericsson K800 launched in July 2006 with a 3.2 megapixel camera and a xenon flash. The K800 was the first Sony Ericsson phone to feature Sony's 'CyberShot' branding.

Orange SPV M5000.

Sony Ericsson K610i.

Sony Ericsson K800i.

Despite having Internet-access capability, it was quite usual for the mobile network operators to restrict access to the public Internet in the early days of handset based mobile data and instead offer their own services within what were known as 'walled gardens'. However, most operators had opened up full access to the public Internet during the early years of 3G – one example of integrating the mobile phone with the Internet was the X-Series products from Three. Launched in November 2006, the X-Series integrated innovative online services such as Skype, Slingbox, Orb and Ebay along with search and Internet features from Google, Microsoft and Yahoo. At the X-Series launch event, Canning Fox, Group Managing Director of Three's parent company Hutchison Whampoa, said: 'This is the Internet as it was meant to be and what people have been waiting for. Mobile broadband is the natural next step for mobile services, extending the full power of the internet to mobile handsets.' The first two handsets to support the full range of X-Series services were the Nokia N73 and Sony Ericsson W950i.

The next big thing in 3G would be the 'dongle', a USB-based cellular radio module that plugged in to a port on a laptop computer to enable instant Internet access. From the launch of 3G services in 2003 and 2004, the UK's five mobile network operators had witnessed steady

Nokia N73.

data traffic growth driven mainly by data cards, however the dongle would soon prove to be a game changer. Laptops were increasingly accessible as mass-market volumes drove prices down significantly and the younger generation joined business people as regular users of connected laptop computers. By 2008, mobile broadband traffic was starting to grow at a phenomenal rate and the demand for ever faster data and larger data bundles resulted in a very competitive market for which the consumer was the winner. This phenomena was often referred to as 'dongle mania' by the industry and mainstream press.

The 3G network architecture is similar to that illustrated in the network architecture diagram in Chapter Three in as much as there was a circuit-switched telephony network in parallel with a data network. The radio access network had evolved and therefore the connections between radio base stations and the core network, along with connections within the core and to external networks, scaled up considerably to support broadband mobile data communications.

Internet access drove demand for mobile data and 3G had to evolve to address this need. The industry therefore specified new standards to offer High Speed Packet Access (HSPA) to provide a best effort, high-speed data downlink channel. Over the following years the 3G downlink and uplink continued to evolve, with downlink capability reaching 21Mbps with the introduction of new data transmission

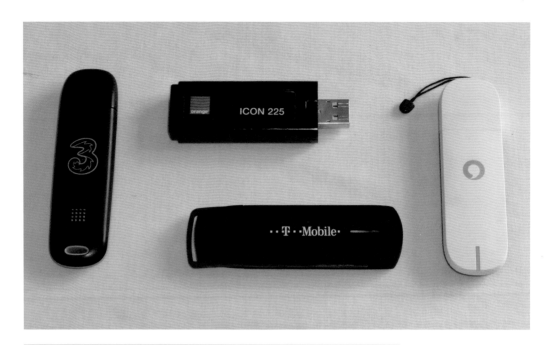

Above: 3G USB dongles.

Left: Huawei E585 Mobile Wi-Fi.

Samsung SGH-Z630 – slim and sleek.

HTC TyTN with QWERTY keyboard.

Nokia N95.

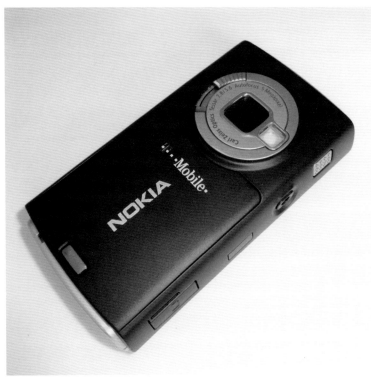

Nokia N95 with
T-Mobile branding.

techniques and methods. These developments are often marketing as 3G+. Mobile Wi-Fi devices which provide local Wi-Fi connectivity while using 3G+ to connect through the cellular network to the Internet became extremely popular, an example is the Huawei E585.

The range of 3G device form factors evolved to include fashionable slim phones, such as the Samsung SGH-Z630, and devices optimised for data input with a QWERTY keyboard, such as the HTC TyTN.

We've seen examples of Nokia N-Series phones previously in this chapter, however one of the best was still to come. The Nokia N95 was an incredibly popular 3G phone and one of the last devices to look like a traditional mobile phone. The evolution that would follow would change the way the mobile phone looked forever.

Chapter Five

Smartphones

By the end of 2006 the mobile industry and its users knew what a smartphone was, what it looked like and what they expected from it. It looked like the Nokia N95, which had been announced in September as an all-in-one multimedia computer. Little did people realise that smartphone design would be dramatically and irrevocably turned on its head in 2007 when a computer company decided to enter the mobile market.

The first mobile or cellular phones were exactly that, devices for making and receiving telephone calls. However, in the early 1990s, alongside the development of the mobile, personal computers were beginning to mature and use software that was increasingly icon and graphic orientated. People were becoming more aware of the Internet thanks to the emergence of the web and small handheld organiser, or personal digital assistant (PDAs), devices were being used as electronic diaries. Inevitably there was a drive to start integrating these.

IBM is credited with taking the first step in that direction when they, along with telecommunications provider BellSouth, jointly announced the Simon Personal Communicator Phone on 8 November 1993. This device combined the features of a PDA with that of a mobile phone, thereby offering users email, a calculator, calendar, clock, phone, address book and notepad – all accessible through a stylus-activated touchscreen.

In Europe, Nokia announced the Nokia 9000 Communicator in 1996 as the world's first all-in-one mobile communications tool for the second-generation GSM networks. This clamshell device had two interfaces: the phone interface, which functioned as a conventional mobile phone, and the communicator interface which could be accessed by opening the phone to reveal a full QWERTY keyboard and large screen. Internally it had 8 MB of memory, of which 4 MB was used for its GEOSTM 3.0 operating system, 2 MB for programs and 2 MB for user data, and it was powered by an Intel 24MHz 386 microprocessor. In addition to all of the basic phone, address book and SMS functions, the Nokia 9000 had a document handling and editing feature called Notes, a fully featured calendar and diary application, a clock showing world time and including an alarm. It could also be connected to a PC via an infrared port through which documents could be exchanged and software downloaded and installed. However, the most important feature was connection to the Internet using Nokia's Smart Messaging for access to email, remote computing and elementary information services.

Nokia 9000 Communicator.

Nokia continued developing its Communicator range culminating in the E90, which was announced in February 2007 as the device for use on 3G networks that 'sets the standard for an uncompromised mobile office experience.'

Nokia E90.

Another innovative product that combined a PDA with a GSM mobile phone and a digital camera was the Orange Videophone, released in 2000. Designed in-house by Orange, the Videophone used Microsoft Windows CE as its operating system and used innovative technology capable of exchanging high-quality video over the very limited data carrying capacity of a GSM network using High Speed Circuit Switched Data (HSCSD). In addition to video, the device also supported email, a calendar that could be remotely synchronised and general Internet access. While this product did not survive in the market for long, it did inspire a new range of Orange smartphones known as the SPV range.

The Orange Sound, Pictures, Video (SPV) range of mobiles were manufactured by the Taiwanese HTC Company and were the first smartphones to be powered by the Microsoft Windows Mobile operating system and intended to bring top-end functionality to the masses. Mini versions of desktop applications such as Outlook, Media Player and Internet Explorer, alongside calendar, task and contact management applications, were all included. The SPV Classic (HTC Canary) was released in 2002 and was further developed as the EDGE-enabled SPV E650 (HTC Vox), released in 2007 as the first device on the UK market powered by Microsoft Windows Mobile 6.0. It also introduced a fully QWERTY keyboard accessed by sliding the device open.

Orange Videophone.

Right: Orange SPV Classic (HTC Canary).

Below: Orange SPV E650 (HTC Vox).

The first mobile phone to be officially marketed as a smartphone was the Ericsson R380 World. Launched in 2000 it was the first GSM mobile to use the EPOC32 operating system, which was originally used in Psion PDA organisers but then developed to become the Symbian Operating System. The R380 adopted a flip design in which a conventional phone keypad partially obscured a large touchscreen. When open, the whole of this screen could then be used for smartphone applications with text entry either being done using a QWERTY keyboard displayed on the touch-sensitive screen or by a handwriting recognition application.

Following the creation of Sony Ericsson in 2001, the R380 was superseded by the Sony Ericsson P800 in 2002 and then the P900 in 2003 – both of which used the Symbian operating system. Work on Symbian started in June 1998 as a joint venture between Nokia, Psion, Ericsson and Motorola, but from June 2008 was taken forward solely by Nokia. Up until 2010 Symbian was the world's most widely used operating system within smartphones, but then rapidly lost ground to competitors to such an extent that by 2013 even Nokia announced they were no longer shipping Symbian-based smartphones. Symbian was also used in a range of specialist smartphones produced by Nokia and primarily intended for the gaming market. The Nokia N-GAGE was launched in 2003 as a combined games console and mobile phone but failed to make an impact on the UK market.

Ericsson R380 World Smartphone.

Sony Ericsson P900.

Nokia N-GAGE.

In 2002, the mobile operator O2 commissioned the design of a range of smartphones marketed under the xda brand. Here the 'x' represented convergence of phone and information while 'da' stood for digital assistant. These devices were manufactured by a range of vendors but mainly HTC. In common with the Orange SPV range, these devices used the Windows Mobile operating system.

Another very important aspect of technology integration that has become an essential feature of a modern-day smartphone is access to Global Positioning System (GPS) satellite navigation. Finnish mobile phone manufacturer Benefon launched the Benefon Esc! in late 1999 as the world's first combined mobile phone and GPS satellite navigator. The Esc! allowed users to trace their position on maps downloaded to the phone and to programme an emergency key which when pressed would send their co-ordinates via SMS text message.

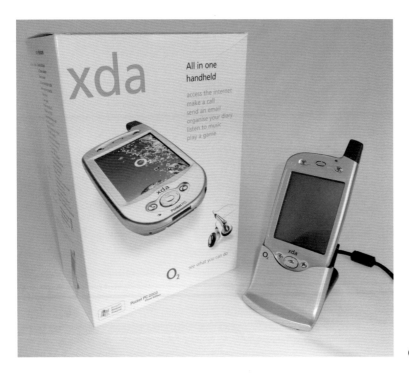

O2 xda.

Inclusion of a navigation capability into smartphones allowed for the development of a new generation of location-based services, which meant that all of the major manufacturers now sought to include GPS in their products. As an example, the W760i was the first of Sony Ericsson's Walkman mobiles to include GPS and came with Google maps preloaded and GPS running performance-monitoring software called Tracker.

Another pioneer of the smartphone was Canadian company Research in Motion (RIM), who specialised in producing products that fully integrated email with a mobile phone – such as the Blackberry 5820, which became available in the UK on Cellnet in 2001. Always favouring the inclusion of a full QWERTY keyboard, the Blackberry not only established itself as a market-leading device for the business user but also became one of the most recognisable and iconic brands of its time. Typical of its products was the Blackberry Curve range, first introduced with the 8300 in 2007 and then extended to the 8900 in 2009, which offered EDGE connectivity. However, by this time there was a more general move towards large touchscreens and so RIM responded with the Blackberry Storm 9500, launched in 2008 and available on Vodafone within the UK. However, it proved to be a commercial failure and prompted a return to the familiar design format with models such as the Blackberry Bold 9700, released in December 2009. Nevertheless, RIM's star had started to wane and from a high point in 2009, and despite rebranding the company as

Sony Ericsson Walkman w760i.

Blackberry in 2013, their global market share of smartphone sales has now fallen below 1 per cent.

Every so often a product enters a market and has such a dramatic impact that it completely transforms that market. The Apple iPhone, launched at the Macworld conference in January 2007, is a classic example of such disruptive technology and lies at the heart of why previously dominant companies such as Nokia and Blackberry have now massively lost their market share of global smartphone sales. Up until the launch of the iPhone, smartphone design was driven by companies that had seen it as a phone first and computer second. However, the Apple design team clearly had a different view and turned conventional wisdom on its head. Whatever your view of Apple, there is no debating the fact that prior to the launch of the iPhone, smartphones looked similar to Nokia or Blackberry products but afterwards all looked like the iPhone.

The iPhone went on sale in the UK in November 2007 on O2 and was in many ways a pretty average 2G phone, but by the end of that year almost 1.4 million had been sold globally. For UK customers there were plenty of technically superior products on the market at that time, but the iPhone offered something quite new. Firstly, the whole design concept was orientated around the touchscreen, for which you used your fingers, not a stylus, and navigated by gestures such as flicking through images and multi-touch controls such as pinch-to-zoom. Secondly, the web browser rendered pages on the small screen much closer to the computer experience. Thirdly, and perhaps

Blackberry Curve 8900.

Blackberry Storm 9500.

Blackberry Bold 9700.

most importantly, Apple provided the App store through which third-party providers could distribute programs to run on the device.

The Apple iPhone was named the best innovation of 2007 by *Time* and was upgraded for connectivity to 3G networks in 2008 with the release of the iPhone 3G and then the 3GS in 2009.

The capabilities and functionality of a smartphone are defined by the software of its operating system. Each smartphone manufacturer had in turn developed their own proprietary system but, on 5 November 2007, the first version of the Android smartphone operating system was released by Google. An important distinction with Android is that it is fully available to any manufacturer to include within their products, which has seen many abandon their own proprietary software in favour of developing Android-based devices. Android is primarily designed for touchscreens with gesture control and is fully integrated with Google's own applications such as Maps, Calendar and Gmail. However, Android also supports third party apps which are made available through the Google Play store. The first smartphone to use the Android operating system was the HTC Dream, released in the UK on 30 October 2008. Today, 80 per cent of smartphones use Android. Typical of these is the Samsung Galaxy S4, which was launched in April 2013 and not only became Samsung's fastest-selling smartphone, with 20 million being sold globally within the first two months, but at that time was also the fastest-selling Android phone too.

The evolution of the smartphone was inevitable given the growth in mobile ownership, our expanding usage of (and reliance on) the web and continual improvements in electronics and software. However, the smartphone has dramatically redefined not only the

Original Apple iPhone from 2007.

Apple iPhone 3GS.

mobile phone itself but also the networks to which they connect, where there has been a clear shift towards offering better support for data. Mobile manufacturers that once dominated the industry, such as Motorola and Nokia, have now lost their market share to competitors predominantly from the Far East. For 2014, assessment of worldwide smartphone sales by Gartner put the top five manufacturers now as Samsung (24.7 per cent), Apple (15.4 per cent), Lenovo (6.5 per cent), Huawei (5.5 per cent) and LG Electronics (4.6 per cent). With regard to the operating system which these smartphones use, Android dominates with 80.7 per cent of the global market share followed by Apple iOS (15.4 per cent), Windows (2.8 per cent) and Blackberry (0.6 per cent). Of the two big app stores, Google Play has now overtaken Apple with 1.43 million apps compared to 1.21 million.

Against this backdrop there is, however, a very important and successful UK company. Cambridge-based ARM Holdings designed the processor used in the Ericsson R380 and today boasts that its designs can be found in 95 per cent of the world's smartphones.

According to Ofcom, the telecommunications regulator, 61 per cent of UK adults now own a smartphone, with the highest level of ownership (88 per cent) being the 16–24 age group. The smartphone story is one in which the mobile phone ceased to be a telephone and morphed into a pocket-sized computer. Today's mobile smartphones and their associated networks are therefore focused on delivering an ever richer and more capable multimedia experience.

Samsung Galaxy S4.

Chapter Six

The Broadband Multimedia Experience

The fourth generation of mobile communications technology, 4G or LTE as it's commonly known within the mobile ecosystem, arrived in the UK on 30 October 2012 with the launch of EE, which was a rebranding of Everything Everywhere. Everything Everywhere had been formed a couple of years earlier by the merger of Orange UK and T-Mobile UK, but EE was to be the dedicated 4G brand whereas Orange and T-Mobile products were 2G and 3G only. Customers on 4G do of course have access to 2G and 3G when outside of a 4G coverage area or when making or receiving a telephone call during the first few years of 4G service – more on this later. Ofcom had granted EE permission to launch 4G services over their existing frequency spectrum in the 1800MHz band and hence their network launched prior to the official 4G auction. The 4G spectrum auction was for new mobile spectrum in the 800MHz and 2600MHz bands and there was no guarantee of a new entrant, however the auction did attract seven bidders. These consisted of the four existing mobile operators along with three other interested parties. Ofcom published the results of the auction and final awards of spectrum on 1 March 2013; there would be a new entrant after all, and this took the number of UK mobile network operators back to five. The total amount raised from the auction was £2,368,273,322 – not an insignificant sum but considerably less than operators had paid for their 3G licences in the year 2000. The financial consequence of the 3G auction on the UK mobile operators impacted their ability to invest in new networks, which in turn meant that the large-scale availability of 3G was delayed by a few years. The more realistically priced 4G auction, alongside Ofcom's approval for EE to start deploying 4G in existing spectrum prior to the auction, ensured that 4G would be rolled out more quickly by all four established UK mobile operators. Vodafone and O2 launched 4G in August 2013, followed by Three in December 2013. The final outcome of the 4G auction is detailed in the following table:

Everything Everywhere Ltd (EE)	Spectrum in the 800MHz band and 2600MHz band for £588,876,000
Hutchison 3G Limited (3)	Spectrum in the 800MHz band for £225,000,000
Niche Spectrum Ventures Limited (BT)	Spectrum in the 2600MHz band for £201,537,179
Telefonica UK Limited (O2)	Spectrum in the 800MHz band for £550,000,000
Vodafone Limited	Spectrum in the 800MHz band and 2600MHz band for £802,860,143

In addition to the 800MHz spectrum which Three purchased in the auction, they also have spectrum in the 1800MHz band which they're using for 4G. This was purchased from EE during 2013. EE had to divest some of its 1800MHz spectrum in accordance with the conditions set by the European Commission for approval of the merger of Orange UK and T-Mobile UK in March 2010. An important difference with a 4G network is that it is designed as a data-only network and therefore, unlike all of its predecessors, does not have any inherent support for standard telephony services and therefore a new technique had to be found to provide 'Voice over LTE' (VoLTE). This new technique is based on an 'IP Multimedia Sub-system' (IMS) which supports generic 'Voice over IP' (VoIP). For the mobile network VoLTE is a specific form of VoIP and uses similar infrastructure to native 'Voice over Wi-Fi' (VoWi-Fi) calling. Both services entered commercial service during 2015 with some of the UK mobile network operators. Prior to VoLTE, any 4G telephone calls used a technique known as 'Circuit Switched Fall-Back' in which the mobile phone is redirected to the 3G network to handle voice telephony, therefore using the legacy circuit-switched network. If 3G coverage isn't available the 4G phone drops back to 2G if the operator supports this mode. Referring back to the network architecture diagram in Chapter Three, the 4G network consists of radio base stations which connect directly to an evolved packet core – there is no network controller as in 2G and 3G and the traditional telephony core network does not exist. The new 4G network has been implemented in parallel with 2G and 3G using the same radio sites and core network sites for the network components.

Prior to the launch of 4G in the UK there was much talk in the press about the UK falling behind as 4G networks had already launched in many other countries. In fact, the 4G standard had been ready since early 2009. In reality the early 4G device portfolio was extremely limited – typically devices were data-only, such as USB data dongles. Additionally mobile Wi-Fi hotspots were available. 4G phones arrived just in time for the commercial launch of 4G services in the UK. EE had six 4G phones available on the day they announced their 4G product range; HTC One XL, Huawei Ascend P1 LTE, Nokia Lumia 820, Nokia Lumia 920, Samsung Galaxy Note 2 LTE and the Samsung Galaxy S3 LTE. A seventh phone, the Apple iPhone 5, was added the following day.

Alcatel 4G LTE USB dongle.

Huawei E5776 Mobile Wi-Fi.

Samsung Galaxy S3 LTE.

Today's 4G networks operate in the 800MHz, 1800MHz and 2600MHz bands, however in time other existing bands at 900MHz and 2100MHz will be used along with new spectrum. The 4G network introduces another new radio interface technology called Orthogonal Frequency Division Multiple Access (OFDMA). Additionally, since the network is a broadband data-only network, its data handling capacity has scaled significantly to support large volumes of multimedia content.

The pace of change isn't slowing. Since EEs 4G launch, Samsung has released three new models of their high-end Galaxy S smartphone, the S4, S5 and S6. Apple has released three new versions of the iPhone: the 5C, 5S and 6, the 6 being available in two different sizes. Other vendors continue to innovate with new devices from Huawei, Nokia and Sony among others.

Previous generations of mobile communications technology introduced alternative form factors for devices such as data cards and USB dongles. This trend continues with 4G as an ever more diverse range of devices become connected to mobile networks. Within mobile communications industry parlance, the term UE is used to describe any device which authenticates with and connects to the network. UE refers to User Equipment and this use of terminology applies to many devices which one wouldn't necessarily consider to be mobile phones. These range from home routers, which offer an alternative to fixed broadband, to connected cameras,

Left: Apple iPhone 5.

Below: Samsung Galaxy S3, S4 & S6 smartphones.

Apple iPhone 6.

tablet PCs, even cars and trains. The possibilities in reality are only limited by our imagination.

The latest developments in 4G have focused on Carrier Aggregation (CA), which is an LTE-Advanced technique that enables several radio channels to be grouped together to offer higher peak and average data rates along with greater system capacity and performance. The first implementation of CA in the UK effectively elevated the UK from its status as a 4G laggard to the home of the fastest 4G city in the world. This was achieved by combining spectrum in the 1800MHz band with spectrum in the 2600MHz band to offer a theoretical peak downlink data rate of 300Mbps, albeit actual user data rates are somewhat lower in a live network environment due to the shared nature of radio capacity. CA capability is currently being rolled out by two UK 4G operators to major population centres. The most recent UK demonstration of CA achieved peak data rates above 400Mbps.

Having established itself as a leading 4G nation, the UK is keen to continue its leadership in mobile communications. While 4G continues to evolve to deliver greater peak and average speeds along with new and innovative services, 5G will be the next big thing. The British government, in partnership with industrial partners and the University of Surrey, has created the 5G Innovation Centre (5GIC). With an initial investment of £35 million, the 5GIC provides the UK

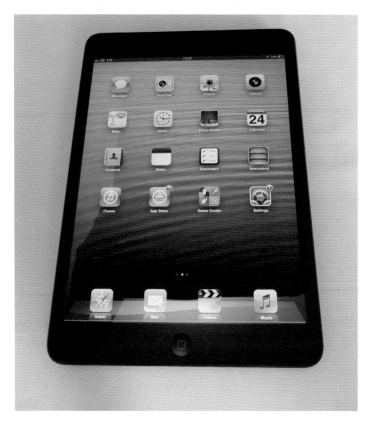

Left: Apple iPad mini with 4G connectivity.

Below: Huawei B593 4G home router.

with a cutting-edge research facility to ensure continued leadership in not just mobile communications but also the evolving Internet of Things.

In summary, the UK has been part of the mobile world for thirty years and what an amazing journey it has been! From analogue car phones and yuppie bricks through to stylish and sleek mobile telephones, and all the way to today's high-end smartphones and increasingly diverse range of applications and services. The famous and pioneering companies from the early days of mobile phones are no longer the leading manufacturers of handsets, in fact many of these organisations don't exist anymore. The introduction of 2G GSM, with its move to digital transmission techniques, ensured a mass market which increased volume and drove down prices, making the mobile phone a practical reality for everyone. Digital techniques enabled ever greater data capability on devices and led to the full integration of the mobile phone with the Internet, changing the way people communicate and access information forever. The 3G UMTS auction stretched the finances of the network operators to almost breaking point and resulted in significant changes to the UK mobile network landscape, and also resulted in significant delays to the rollout and adoption of 3G services. The launch of the Apple iPhone in 2007 fundamentally changed the way we thought about the mobile phone and rewrote the design book for what a high-end smartphone should be. Moving to 4G LTE was perceived to be late in the UK. However, due to sensible regulatory management and a realistically priced spectrum auction, we've seen the UK progress to become a world leader in 4G networks and services.

Looking to the future, research into 5G is ongoing. What this will actually be capable of will become clearer over the next couple of years however one thing is for sure – it'll be better again with vastly improved performance, capacity and coverage. This will therefore drive the ongoing mobile communications eco-system and underpin the evolving Internet of Things.

What will mobile phones look like in another thirty years? The authors have been in this industry for too long to tackle this question given the many failed predictions that litter the subject's history. However, one thing is for certain, within thirty years just about everything that can be connected will be, and this will have a significant impact on the way we live our lives – just as we have witnessed in the past thirty years.

Acknowledgements

The authors are grateful for the kind co-operation and assistance from many colleagues and companies, with particular thanks being given to: Stephen Temple, Vodafone, Mike Short, Mike Pinches, Julian Divett, John Faulkner, Matt Chatterley, University of Salford and the Engineering and Physical Sciences Research Council.